Lewis Latimer

The Man Behind a Better Light Bulb

PEBBLE
a capstone imprint

Little Explorer is published by Pebble, an imprint of Capstone
1710 Roe Crest Drive,
North Mankato, Minnesota 56003
www.capstonepub.com

The name of the Smithsonian Institution and the sunburst logo are registered trademarks of the Smithsonian Institution. For more information, please visit www.si.edu.

Library of Congress Cataloging-in-Publication data is available on the Library of Congress website.
ISBN 978-1-9771-1411-2 (library binding)
ISBN 978-1-9771-1786-1 (paperback)
ISBN 978-1-9771-1415-0 (eBook PDF)
Summary: Gives facts about Lewis Latimer, his life, and how his invention of the carbon filament led to more affordable and longer-lasting light bulbs.

Image Credits
Alamy: Old Paper Studios, 6, Robert K. Chin–Storefronts, 29; AP: Hearst Connecticut Media/Cathy Zuraw, 28; Boston Athenæum: 10; Courtesy of the Queens Borough Public Library, Long Island Division, Lewis H. Latimer Collection: cover (top), 5, 11, 14, 15 (top), 17, 19, 22, 23, 24–25, 26, 27; Library of Congress: 9 (bottom), 13 (top), 18; National Archives and Records Administration: 13 (bottom); Newscom: World History Archive, 16; North Wind Picture Archives: 4 (left), 15 (bottom), 20; Shutterstock: Fenton One, 4 (right); Smithsonian Institution: National Museum of American History, 21 (bottom); U.S. Naval History and Heritage Command: Louis H. Smaus, 9 (top); U.S. Patent and Trademark Office: 12; Wikimedia: Daderot, cover (bottom), 21 (top)

Design Elements by Shutterstock

Editorial Credits
Editor: Michelle Parkin; Designer: Sarah Bennett; Media Researcher: Svetlana Zhurkin; Production Specialist: Tori Abraham

Our very special thanks to Emma Grahn, Spark!Lab Manager, Lemelson Center for the Study of Invention and Innovation, National Museum of American History. Capstone would also like to thank Kealy Gordon, Product Development Manager, and the following at Smithsonian Enterprises: Ellen Nanney, Licensing Manager; Brigid Ferraro, Vice President, Education and Consumer Products; and Carol LeBlanc, Senior Vice President, Education and Consumer Products.

All internet sites appearing in the back matter were available and accurate when this book was sent to press.

Printed in the United States of America.
PA99

TABLE OF CONTENTS

Bold words are in the glossary.

LIGHTING THE WORLD

When it gets dark, we turn on a light. But 150 years ago, it wasn't as easy as flipping a switch. Most people used candles to light their homes. Electric light bulbs were just being invented.

Before light bulbs, people used candles to see at night.

Lewis Latimer

Lewis Latimer helped change that. He was an **engineer** and **inventor**. Inventors are people who come up with new ideas. Latimer invented a new kind of light bulb. It lasted longer than other light bulbs during that time.

EARLY LIFE

Lewis was born on September 4, 1848.
He was the youngest of four children.
His family lived in Chelsea, Massachusetts.
This small city is just outside of Boston.

Chelsea, Massachusetts in the mid-1800s

Lewis was a good student. He loved to read and draw. When Lewis wasn't at school, he worked with his father. At first Lewis's father ran a barbershop. Then he had a business putting up wallpaper. Lewis helped with both.

GEORGE LATIMER

Lewis's parents, George and Rebecca, were enslaved people who ran away. They had escaped their **slaveholders** in Virginia and moved to Massachusetts. George Latimer was arrested for running away in 1842. His former slaveholder came to Boston to take George back. But George fought him in court. Friends raised money to buy George's freedom. Eventually George Latimer became a free man.

When Lewis was 10 years old, his father left the family. His mother struggled to support Lewis and his siblings. Lewis tried to help. When Lewis was 13, he got a job working in a lawyer's office.

The **Civil War** began in 1861. A few years later, Lewis wanted to join the **Union** Navy and fight in the war. But he was too young. Lewis lied about his age to get into the Navy.

The Civil War was fought between northern and southern states. It lasted from 1861 to 1865.

USS Massasoit

Lewis served on the Navy ship USS Massasoit.

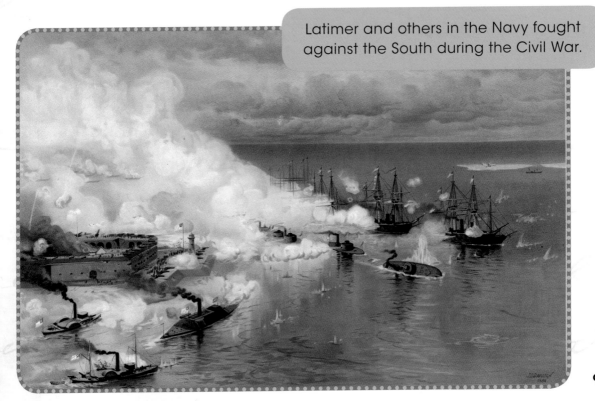

Latimer and others in the Navy fought against the South during the Civil War.

LEARNING A NEW SKILL

After leaving the Navy, Latimer got a job at a law **firm** called Crosby, Halstead, and Gould in Boston, Massachusetts. The firm helped inventors apply for **patents**. A patent protects an inventor's work. It keeps other people from copying the inventor's ideas and making money from them.

law firm office sign

The Crosby, Halstead, and Gould law firm in Boston

Latimer filed papers and ran errands. But he wanted to be a **draftsman**. Draftsmen make detailed drawings of inventions. The drawings were needed to apply for patents. Latimer read books about drafting. He taught himself to use drafting tools such as compasses and rulers. He watched other draftsmen at the office. Eventually Latimer became a skilled draftsman.

drafting tools

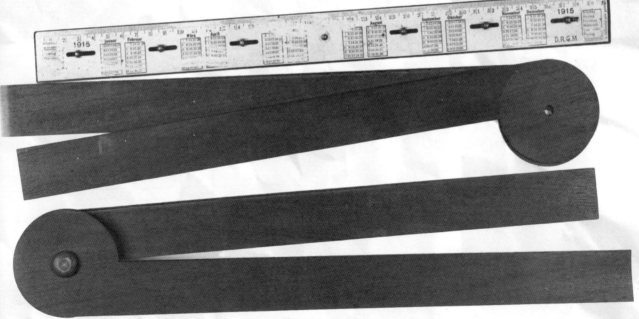

Latimer's bosses noticed his talent. After one of the draftsmen left the company, Latimer got his job. In his spare time, Latimer came up with his own inventions. In 1874, Latimer got his first patent for an invention. It was a new toilet system for trains. The toilet emptied through a trapdoor in the train's floor.

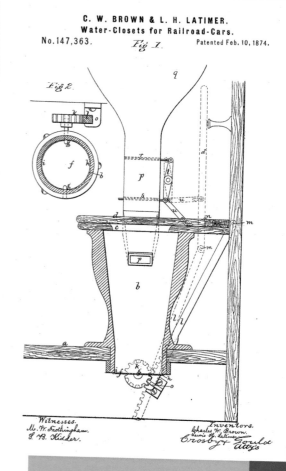

Latimer's patent for a toilet system on trains

Two years later Latimer helped another famous inventor—Alexander Graham Bell. Bell invented the telephone. Latimer made the technical drawings needed for Bell's patent.

Other inventors wanted to be the first to patent the telephone. Thanks to Latimer, Bell turned in his paperwork first. Bell beat another inventor by a few hours.

Alexander Graham Bell

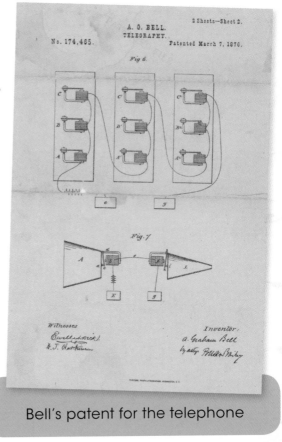

Bell's patent for the telephone

Latimer received 7 patents throughout his life.

NEW OPPORTUNITIES

Latimer left his job in 1878. He had learned a lot. He wanted to take his skills somewhere new. His sister lived in Bridgeport, Connecticut. She suggested that he move there.

Latimer and his wife, Mary, arrived in Bridgeport in 1879. He found a job making technical drawings. One day Latimer met Hiram Maxim. Maxim owned the U.S. Electric Lighting Company. It made electric lights. Maxim was impressed with Latimer's work. He offered Latimer a job as a draftsman.

Mary Latimer in 1882

HIRAM MAXIM

Hiram Maxim was also an inventor. As a teenager, he invented a mousetrap. He patented a hair-curling iron in 1866. Later Maxim invented headlights for trains. His most famous invention was the automatic machine gun. He invented it in 1884.

Hiram Maxim invented the Maxim field gun.

ELECTRIC LIGHT

Early light bulbs worked by sending electricity through a thin **filament**. The electricity made the filament glow. But the filament got very hot. It could melt or even catch fire. This could be dangerous.

A drawing of an electric light bulb from the 1860s

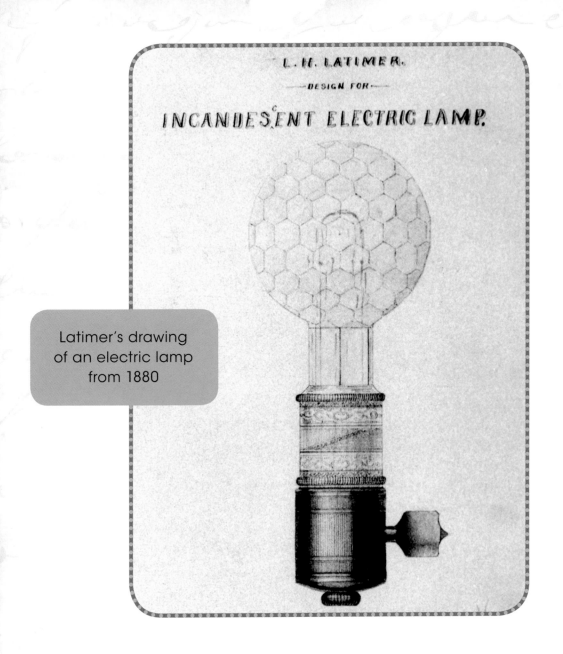

Latimer's drawing of an electric lamp from 1880

Many people tried building a better light bulb. It was a race to see who would invent it first! Latimer worked with others on Maxim's team to try creating the new bulb.

CARBON FILAMENTS

Thomas Edison invented one of the first practical light bulbs. His company competed with Maxim's company. Edison's light bulbs used **carbon** filaments. They only lasted for a few days. Latimer wanted to do better.

Thomas Edison

ELECTRIC LIGHTS THROUGH TIME

1803—Chemist Humphry Davy makes the first electric light.

1850—German scientist Heinrich Geissler sends an electric current through a glass tube filled with gas. The glass tube glows.

1878—Physicist Joseph Swan makes a glass light bulb.

1879—Inventor Thomas Edison makes a light bulb with a carbon filament.

Latimer tested different filaments. Finally he created a filament that lasted a long time. His invention made electric light less expensive. People could afford to use it in their homes.

Latimer's drawing of a light bulb in 1880

1882—Lewis Latimer patents a new way of making filaments.

1904—Carbon filaments are replaced with brighter tungsten filaments.

1955—Scientists Elmer Fridrich and Emmet Wiley make a light bulb filled with iodine gas. The gas makes tungsten filaments burn longer.

2000s—Countries start to ban incandescent light bulbs. They are replaced with light bulbs that use less energy.

TECHNICAL EXPERT

Electric lighting along a street in London, England

Maxim's company sold Latimer's filaments to people around the world. Companies ordered new light systems for their buildings. Maxim sent Latimer to help set them up. Latimer traveled all around the country. He even went to Canada and Great Britain.

A light bulb with Latimer's filament

Maxim put Latimer in charge of making the light bulbs. Latimer managed a team of 40 workers. Latimer kept inventing other things.

Arc lamps were one of the first kinds of electric lights. Latimer designed a new holder for the lamp.

Latimer drew parts of an arc lamp.

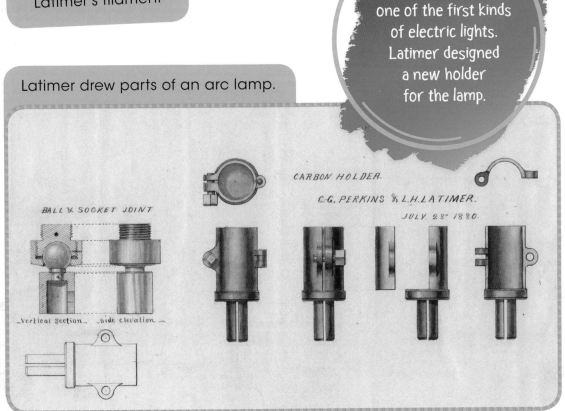

BALL & SOCKET JOINT

CARBON HOLDER.

C.G. PERKINS & L.H. LATIMER.

JULY. 23 1880.

Vertical Section   Side elevation

EDISON PIONEERS

Thomas Edison wanted Latimer to work for him. In 1885, Latimer left Maxim and the U.S. Electric Lighting Company to work for Edison. Latimer was the only black engineer on Edison's team. Years later, Latimer and other members of the team formed a group called the Edison Pioneers.

Latimer and other members of the Edison Pioneers

Latimer

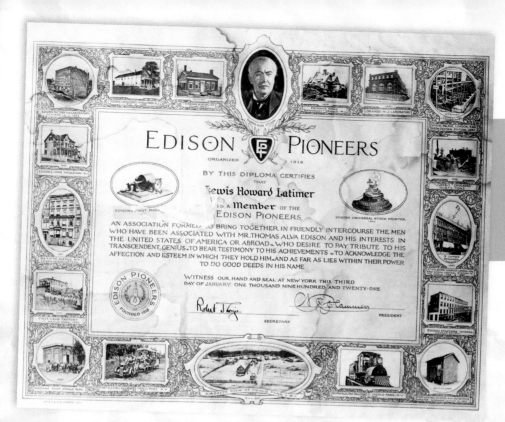

Latimer's certificate for being a member of the Edison Pioneers

Latimer was an expert in electric lighting and patents. He helped other inventors with their patents. But his main job was to protect Edison's inventions. Latimer looked at other companies' lighting systems. He made sure they weren't using Edison's ideas.

In 1890 Latimer wrote the first detailed book about electric lights.

Latimer began to study law. It was useful at his job. Latimer toured the competition's factories. Latimer had to prove they were copying Edison's inventions. He collected information that could be used in court.

In 1889, Edison sued the U.S. Electric Lighting Company for using his inventions. By then, Maxim had sold the company and moved on. Latimer helped Edison win the case.

Edison's legal team

Latimer

LATER LIFE

There were not a lot of black engineers in Latimer's time. They were mostly white men. Latimer supported **equal rights** for black Americans. He felt that a person's hard work and success should earn them respect. The color of a person's skin should not matter.

Lewis Latimer at age 70

Latimer left Edison's company in 1911. He continued to help inventors with patents as a patent consultant. Eventually Latimer started losing his eyesight. He retired in 1924. Lewis Latimer died on December 11, 1928. He was 80 years old.

Latimer with his family at home in 1920

LATIMER'S LEGACY

Lewis Latimer was an important black inventor. He grew up poor, but he worked hard. Latimer always wanted to learn more.

Thomas Edison and Hiram Maxim were famous when they were alive. But Latimer was not. He was not recognized for his electric light inventions until years after his death. Latimer's work helped change the world. His inventions led to the lights we use today.

A statue of Lewis Latimer is in Bridgeport, Connecticut.

MARGARET E. MORTON GOVERNMENT CENTER

Latimer's house was turned into a museum after he died. The Lewis H. Latimer House Museum is located in Flushing, New York.

"We create our future, by well improving present opportunities: however few and small they be."

—Lewis Latimer

GLOSSARY

carbon (KAHR-buhn)—a chemical element found in all living things and in coal

Civil War (SIV-il WOR)—a war between states in the North and South that led to the end of slavery in the United States

draftsman (DRAFT-mahn)—a person who draws detailed drawings of machinery or structures

engineer (en-juh-NEER)—someone trained to design and build machines, vehicles, bridges, roads, and other structures

equal rights (EE-kwuhl RITEZ)—everyone is treated the same under the law

filament (FI-luh-muhnt)—a thin wire inside a light bulb that is heated with electricity to produce light

firm (FURM)—a business or company

inventor (in-VENT-ohr)—someone who thinks up and creates something new

patent (PAT-uhnt)—a legal document giving someone the sole rights to make or sell a product

slaveholder (SLAYV-hohl-duhr)—someone who forced enslaved people to work without pay and under brutal conditions

Union (YOON-yuhn)—the northern states that fought against the southern states during the Civil War

CRITICAL THINKING QUESTIONS

1. Why do you think Lewis's parents went to Massachusetts after they escaped Virginia?

2. Why is it important for inventors to apply for patents?

3. How do you think the invention of the light bulb changed the way people lived?

READ MORE

Marsico, Katie. *Lewis Howard Latimer.* Ann Arbor, MI: Cherry Lake Publishing, 2018.

Mattern, Joanne. *Lightbulbs.* New York: Children's Press, 2016.

Raatma, Lucia. *Thomas Edison: The Man Behind the Light Bulb.* North Mankato, MN: Pebble, 2020.

INTERNET SITES

Facts About Lewis Latimer
https://lemelson.mit.edu/resources/lewis-h-latimer

National Inventors Hall of Fame
https://www.invent.org/inductees/lewis-latimer

INDEX